Northern Illinois University Shooting
DeKalb, Illinois
February 14, 2008

Reported by: Hollis Stambaugh

This is Report 167 of Investigation and Analysis of Major Fire Incidents and USFA's Technical Report Series Project conducted by TriData, a Division of System Planning Corporation, under Contract (GS-10-F0350M/HSFEEM-05-A-0363) to the DHS/United States Fire Administration, and is available from the USFA Web page at http://www.usfa.dhs.gov

Department of Homeland Security
United States Fire Administration
National Fire Data Center

U.S. Fire Administration
Mission Statement

We provide National leadership to foster a solid foundation for local fire and emergency services for prevention, preparedness and response.

ACKNOWLEDGMENTS

The research for this report was greatly facilitated by the generous cooperation and support of the following individuals:

Bruce Harrison	Fire Chief, City of DeKalb Fire Department
Lanny Russell	Retired Fire Chief, City of DeKalb Fire Department
Traci LeMay	Administrative Assistant, City of DeKalb Fire Department
Melanie Magara	Assistant Vice President for Public Affairs, Northern Illinois University
Dennis Miller	DeKalb County Coroner and Emergency Management Coordinator

The USFA and the author of this report also would like to extend our gratitude to Northern Illinois University (NIU) President, Dr. John Peters, and to Virginia Polytechnic and State University (Virginia Tech) in Blacksburg, Virginia. The information that Virginia Tech shared with NIU and others concerning the tragedy at the Virginia Tech campus has significantly contributed to improved campus emergency preparedness at schools across the country, including at NIU. We also honor the victims of the February 14, 2008, assault and extend our sincerest wishes for the continued healing of the families and the campus community.

Remembering the victims of February 14, 2008.

TABLE OF CONTENTS

CHAPTER 1: FIRE, RESCUE, SECURITY, AND EMERGENCY PREPAREDNESS AT NORTHERN ILLINOIS UNIVERSITY 1
    Fire and Rescue 1
    Automatic Aid and Mutual Aid 2
    Campus Security and Crime Statistics 3
    Emergency Preparedness 4
    State of Illinois 5

CHAPTER 2: BRIEF HISTORY OF THE NORTHERN ILLINOIS UNIVERSITY SHOOTER AND PARALLELS TO THE VIRGINIA TECH SHOOTER 7

CHAPTER 3: THE INCIDENT AND MULTIAGENCY RESPONSE WITH TIMELINE 11
    First Actions 11
    Triage and Treatment 13
    Staging 17
    Timeline 22
    Additional Resources 23
    Campus Alert and Public Information 24

CHAPTER 4: HOSPITAL AND CORONER'S OFFICE RESPONSE 28
    Arriving Families and Information 31
    Second Wave of Patients 32

CHAPTER 5: AFTERMATH AND HEALING THE WOUNDS 33

CHAPTER 6: FINDINGS AND LESSONS 37

LIST OF FIGURES

Figure 1: Comparison of the VA Tech and NIU Shooters 10

Figure 2: Tactical Police Forces Rush to the Scene 12

Figure 3: Status of Victims at Cole Hall by Order of Transport 14

Figure 4: Paramedics Rush a Victim to a Waiting Ambulance 15

Figure 5: Diagram of Triage Locations 16

Figure 6: Scene Outside Cole Hall 17

Figure 7: Fire Department Staging Area 18

Figure 8: Medics Had to Walk a Distance From the Triage Sites to the Ambulances 18

Figure 9: Status and Disposition of Patients 19

Figure 10: Law Enforcement Personnel Guard Cole Hall the Evening of the Shooting 20

Figure 11: Placement of Operations on Campus 21

Figure 12: Timeline of Events (3 p.m. February 14 – 12 a.m. February 15) 22

Figure 13: NIU President, Dr. Peters, at the Second News Conference 26

Figure 14: Dr. Michael Kulisz Speaks to the Media at Kishwaukee Community Hospital 30

Figure 15: Hospitals Involved in Treating the Victims 31

Figure 16: Mourners Participate in Candlelight Vigil 33

Figure 17: Crosses Erected Near the Site of the Shooting 34

Figure 18: A View of the Packed Convocation Center During Memorial Service 34

Figure 19: A Demonstration of Hope for the Future 40

BACKGROUND

On February 14, 2008, less than 1 year after a senior at the Virginia Polytechnic Institute and State University (Virginia Tech) murdered 32 people and committed suicide, the campus community at Northern Illinois University (NIU), in DeKalb, Illinois, faced a similar horror. A former NIU graduate student walked onto the stage of a large lecture hall and began firing on startled students and faculty. The shooter, a 28-year old male, had a history of mental illness. He shot and killed 5 students and wounded 18, some critically. His suicide at the end of the brief attack brought the number of deaths to 6.

The building where the shooting occurred, Cole Hall, is centrally located in the interior of the campus and is directly across from a concentration of dormitories identified as Neptune East, West, Central, and North. Cole Hall contains two large lecture halls for large group classes.

All of the injured who were transported were taken to Kishwaukee Community Hospital, the only hospital nearby. Several of the most seriously injured were then transferred to five other hospitals in the region—four via helicopter and one via ground ambulance. A close examination of how the emergency medical and hospital services were carried out reveals that the right decisions and actions were taken during triage and treatment, lives were saved, and no one was hurt in the process of providing emergency medical services (EMS) to the victims, transporting them, or safeguarding the rest of the campus immediately after the murders.

The City of DeKalb Fire Department, the NIU Department of Public Safety, the hospital, and other mutual-aid responders were prepared. They had practiced emergency drills together and coordinated their planning. They were familiar with the Incident Command System (ICS) and had formally incorporated its use in their plans. The fire/EMS, university police, and university events management partners had worked together frequently in planned and unplanned events, so Command and control procedures were well practiced. They also had studied the official report[1] on the Virginia Tech shootings and had integrated the lessons learned enumerated in that report into the university's and the City of DeKalb's emergency response plans, especially from the chapters that reported on the law enforcement and EMS response to that April 16, 2007 incident. The value of that report, their training, and their joint planning was apparent in the excellent response to Cole Hall.

The DeKalb Fire Department has stated they hope that what they discovered from their internal debriefings and reports can add to the lessons that were documented from Virginia Tech so that the body of experience can expand to include this most recent tragedy and help other universities, law enforcement agencies, and fire departments as the Virginia Tech report helped them. The U.S. Fire Administration (USFA) is pleased to enable the sharing of information from the NIU shooting with emergency response organizations nationwide.

---

[1] Mass Shootings at Virginia Tech–Report of the Review Panel, August 2007, Richmond, Virginia.

CHAPTER 1: FIRE, RESCUE, SECURITY, AND EMERGENCY PREPAREDNESS AT NORTHERN ILLINOIS UNIVERSITY

Northern Illinois University (NIU) is situated 65 miles due west of Chicago. The university's 755-acre main campus is in the city of DeKalb, which has a population of 43,714, not including the oncampus housing population of 5,900. NIU consists of seven colleges offering undergraduate, graduate, and doctoral degrees. Total enrollment is approximately 25,000 students, over 90 percent of whom are located at the main campus. The remainder of the student body attends classes at NIU's satellite centers in nearby communities. Nearly all of the students are from Illinois, though approximately eight percent are from other States and countries.

FIRE AND RESCUE

Fire and rescue services at NIU are provided by the City of DeKalb Fire Department; the campus is part of their response area. All regular fire department services are available to NIU. The DeKalb Fire Department employs 58 career firefighters and 2 civilian employees. There are 3 fire stations from which the department provides service to approximately 50,000 people in a 70-square-mile service area that includes the City of DeKalb, Northern Illinois University, and the DeKalb Fire Protection District (an unincorporated area outside the city). Additionally, the fire department provides paramedic-level ambulance service to the Cortland Fire Protection District under a special service contract.

The DeKalb Fire Department provides a full range of emergency services: fire suppression, rescue, emergency medical services (EMS), hazardous materials first response, fire prevention, and public education. There are two divisions: administration and operations. Under the first division, there is the Fire Chief, one Assistant Chief, one Training Battalion Chief, one Lieutenant, one administrative assistant, and one office associate. Personnel in this division work a 37-1/2-hour workweek. The Operations Division functions with a 3-platoon system scheduled as 24 hours on duty and 48 hours off duty. Each shift has 18 fire personnel, a Battalion Chief who serves as shift commander, one Captain, and three Lieutenants operating as Company Officers (COs), each responsible for one of the fire stations. During periods of normal call volume, the department's response time from receipt of an emergency request to the arrival of emergency apparatus on scene is 3 to 5 minutes.

The DeKalb Fire Department enjoys an excellent working relationship with the NIU Department of Public Safety. The two departments participate in joint training and other mutual preparedness activities. Other emergency service organizations, such as the City of DeKalb Police Department, Illinois State Police, local hospitals, DeKalb County Emergency Management, and the DeKalb County Sheriff, are involved in joint exercises and emergency response planning sessions.

In recent years, the emergency services departments serving NIU developed and now routinely use Incident Action Plans (IAPs) to structure their organizational response to scheduled events and to document the evolution of response activities and assignments during an incident. The IAPs follow the National Incident Management System (NIMS), which they have adopted as the model for responding to incidents and supporting those efforts.

The DeKalb Fire Department continues to develop a close-working relationship with campus officials and departments at NIU. When the Governor of Virginia released the final report of his Virginia Tech Review Panel, the fire department downloaded a copy of the report, and carefully studied the chapters related to the services they provide to NIU. In particular, they evaluated the chapter on the EMS response to ensure that they would be prepared if a similar event were to occur at NIU. Four-and-a-half months after the Virginia Tech report was published, the fire department had to apply what it had learned from that tragedy.

AUTOMATIC AID AND MUTUAL AID

The fire department has an automatic-aid agreement with the Sycamore Fire Department to better serve the residents of the area between DeKalb and Sycamore, a community located a few miles directly northeast of the City of DeKalb along Route 23. An automatic-aid request from the Sycamore Fire Department brings a City of DeKalb truck company response, while Sycamore sends an engine company to the DeKalb Fire Department's automatic-aid requests. Both communities send a single ambulance to any motor vehicle accident at certain, designated intersections.

Mutual aid also is available through a Mutual Aid Box Alarm System (MABAS). MABAS is a State and interstate mutual-aid system, which has been in existence since the late 1960s. Pre-September 11, MABAS was heavily rooted throughout northern Illinois. Since September 11, MABAS has rapidly grown throughout the State of Illinois and Wisconsin, and parts of Indiana, Iowa, and Missouri. Day-to-day MABAS extra alarms are systematically designed to provide speedy response of emergency resources to the stricken community during an emergency. Declarations of disasters are not required for routine MABAS system activations. Today, MABAS includes approximately 1,100 of the State's 1,200 fire departments organized within 62 divisions. MABAS divisions geographically span an area from Lake Michigan to Iowa's border and south almost into Kentucky. Eight Wisconsin divisions also share MABAS with their Illinois counterparts. The cities of Chicago, St. Louis, and Milwaukee are MABAS member agencies.

MABAS includes over 37,000 of Illinois' 40,000 firefighters who staff emergency response units including more than 1,300 fire stations, 1,800 engine companies, 389 ladder trucks, 831 ambulances (mostly paramedic capable), 278 heavy rescue squads, and 647 water tankers. Fire/EMS reserve (back-up) units account for more than 1,000 additional emergency vehicles. The system also offers specialized operations teams for hazardous materials, underwater rescue/recovery, technical rescue, and incident management teams. Certified fire investigators can be "packaged" as teams for larger incidents requiring complicated and time-consuming investigations.

MABAS is a unique organization in that every MABAS participant agency has signed the same contract with their 1,100 plus counterpart MABAS agencies. Participating agencies agree to standards of operation, Incident Command, minimal equipment staffing, safety, and onscene terminology. All MABAS agencies operate on a common radio frequency (IFERN) and are activated for response through predesigned "run" cards that each participating agency designs and tailors to meet their local risk need. MABAS also provides mutual aid station coverage to a stricken community when their fire/EMS resources are committed to an incident for an extended period. The DeKalb Fire Department belongs to Division 6 of the area's MABAS, and includes 13 participating and 4 associate fire departments as follows:

Participating (Primary) Departments

DeKalb Fire Department - full career

Sycamore Fire Department - combination

Cortland Fire Protection District - volunteer

Genoa-Kingston Fire Protection District - volunteer

Hinckley Fire Protection District - volunteer

Lee Fire Protection District - volunteer

Malta Fire Protection District - volunteer

Shabbona Fire Protection District - volunteer

Genoa-Kingston Rescue Squad - volunteer

Kirkland Fire Protection District - volunteer

Leland Fire Protection District - volunteer

Somonauk Fire Protection District - volunteer

Waterman Fire Protection District – volunteer

Associate (Primary in Other MABAS Divisions) Departments

Maple Park Fire Protection District - volunteer

Monroe Center Fire Protection District - volunteer

Rochelle Fire Department - combination

Sandwich Fire Protection District - volunteer

CAMPUS SECURITY AND CRIME STATISTICS

The university has a strong commitment to public safety. Law enforcement and security services are vested with the Department of Public Safety, which is led by the police chief. The department emphasizes community policing and works to build relationships with the students, faculty, and staff so that problems are identified early and resolved. The department views their mission as one that helps the campus community learn to keep themselves safe, thus reducing opportunities for crime. The university community is encouraged to cooperate with the police in creating a safe environment and reporting crimes, hazards, or suspicious activities.

The NIU police force includes 60 sworn officers. They have the authority to investigate crimes, make arrests, and help with emergencies. Officers patrol the campus on a 24-hour basis, and the residence halls are established as Community Safety Centers with access control. NIU police are rather unique; all officers are now or are becoming certified as Emergency Medical Technicians (EMTs). That training significantly enhances the services that officers can provide during emergencies and was an important factor during the immediate response to the assault at Cole Hall. More details about the police response are covered in Chapter 3.

In accordance with the "Clery Act"[2], a Federal law, which requires institutions of higher learning to document their crime incident data, NIU's Legal Services section, in cooperation with the Department of Police and Public Safety, prepares an annual security report that shows crime statistics for each NIU campus (Main Campus, Hoffman Estates, Naperville, Rockford, Lorado Taft Field). The most recent of these reports includes data from the year 2007. For the main campus, liquor violations were the prevalent offense, typical for a community with a high concentration of young people. These offenses were largely handled with disciplinary referrals, though some of the more serious cases brought arrests. Burglaries were the next most frequent type of incident, followed by (forcible) sex offenses.

The NIU police have instituted many programs to address the leading types of crime at NIU. For example, they provide comprehensive crime prevention education information on the university's security policies, how to obtain immediate assistance and report a crime (including confidential reporting procedures); how warnings would be authorized and distributed, access control of buildings and grounds; and a special section on safety in the residence halls; and so forth. Each dormitory has a Community Safety Center to which an NIU police officer is assigned.

The university's Division of Student Affairs oversees an NIU Interpersonal Violence Response Team (IVRT). Eight NIU offices, including the Department of Police and Public Safety, make up the team. Victims of sexual assaults, domestic violence, and stalking can confidentially report their situation to any of the IVRT member departments for action.

EMERGENCY PREPAREDNESS

NIU is part of an incident management team known as the Communiversity Incident Management Team: a coalition of campus and local agencies with a stated mission of "Seeking Solutions and Solving Problems through Collaboration." The Communiversity Incident Management Team has both cooperating and participating members. The cooperating members include the NIU Department of Public Safety, the City of DeKalb Fire Department, and the NIU Convocation Center. The participating members of the Communiversity Incident Management Team are

- DeKalb County Sheriff's Office.
- Sycamore Police Department.
- City of DeKalb Police Department.
- City of DeKalb Public Works.
- NIU Athletics.
- Kishwaukee Hospital.
- MABAS Division 6.
- DeKalb County Emergency Management.

---

[2] The Federal Campus Sex Crimes Prevention Act, October 28, 2002.

The Incident Management Team (IMT) developed an IAP which included mass casualty and emergency communications components, and other documents related to emergency preparedness. They organized and conducted special drills to test procedures and made adjustments to their plans that reflected the results of their field exercises.

NIU's multicasualty incident (MCI) plan had been reevaluated in the wake of the April 2007 shootings at Virginia Tech. The day after that tragedy, the NIU President communicated to the university community his intentions to examine the university's emergency plans and preparedness in light of the lessons learned from the Virginia Tech murders. The university convened a meeting of their emergency management team and conducted a "verbal simulation"[3] of the Virginia Tech shootings. The NIU President's letter also signaled the university's intention to hold a series of drills to test and refine their disaster communication and response plans.

Campus and local emergency response agencies practiced the university's MCI plan on October 10, 2007. The NIU Department of Public Safety and the DeKalb Fire Department led a mock mass casualty exercise to practice staging and managing resources. The regional MABAS radio system was tested to ascertain whether a predetermined number of ambulances would in fact be available and accessible during a mass casualty operation. The drill included mutual-aid ambulances, the County Health Department, Kishwaukee Hospital, and air transport (helicopter) support.

Portions of those plans were tested during an actual situation, which occurred on December 8, 2007. On that day, a message was discovered in a dormitory bathroom which threatened that "things will change most hastily" as the semester drew to a close. The message reportedly included a racial slur, underlined letters that spelled out WATCH, and included a question: "What time? The VA Tech shooters messed up with having only one shooter..."[4]

Officials closed the campus and postponed final exams after learning of the message. NIU's Office of Public Affairs used multiple channels of communication to notify everyone on campus about the threat and that the campus was being closed. According to an NIU spokesperson, that system worked well and no major changes to their alert and notification procedures were found to be necessary. NIU followed most of the same procedures on February 14, 2008, as they had 2 months earlier during the threat.

STATE OF ILLINOIS

The State of Illinois established a Campus Security Task Force (CSTF) 2 weeks after the Virginia Tech shootings. The group included leaders in education, law enforcement, mental health, and public safety. Their mission was to identify training needs and provide programs to help security officials at the State's many campuses. CSTF studied campus security issues and is charged with developing procedures to improve safety. For example, Illinois created a Campus Security Enhancement Grant Program, which the Task Force recommended. The grants can be used for response training, violence prevention, and emergency communications equipment and messaging systems, exercises and plans, and awareness programs. In conjunction with the Illinois Emergency Management Agency (IEMA), CSTF developed "All-Hazard Emergency Planning for Colleges and Universities"—a two-phase training initiative.

---

[3] "NIU had tested response plan," Daily Chronicle, February 15, 2008.

[4] Ibid

Phase I provides a 5-day training course for senior college and university administrators, urges them to revise their Emergency Operations Plans (EOPs) in keeping with documented "best practices," identifies hazards common to campuses, and offers information about planning teams and EOP development and testing. The five components of Phase I are

1. Introduction to Campus Emergency Planning Process.
2. Responding Using the Incident Command System Planning Corporation.
3. Critical Incident Stress Management.
4. Crisis Communications.
5. Virginia Tech Tragedy–Lesson Learned and Key Recommendations.

Phase II curricula content will be derived from an evaluation of Phase I and the subject matter that participants indicate is needed.

The Task Force distributed interoperable radios to over 70 colleges and provided campus security awareness training courses to over 95 campuses statewide. NIU was an active member of this State task force, and a recipient of radios and the training prior to the shooting at Cole Hall, and those assets contributed to the response on February 14.

Finally, on April 18, 2008, 2 months after the slayings at NIU, the Task Force published its report, *State of Illinois Campus Security Task Force Report*, to the Governor. The report contains scores of recommendations and specific guidance on such topics as threat assessment, use of and compliance with NIMS, volunteer management, training, emergency drills, and others. There also is a five-page chapter entitled, "Observations and Lessons Learned, February 14, 2008, Shooting Incident at Northern Illinois University." The chapter describes the background on NIU emergency training, simulations, and capabilities, and then presents some information about the NIU police response, emergency communications, public information, and recovery. Several of the findings under Incident Command, however, are not completely in line with NIMS standards, for example, stating that the best plan is for the Chief of Department to remain at the scene. The nationally applied rudiments of NIMS call for the chiefs of the first responder departments to locate where all the key stakeholder departments can coordinate information, resource requests, Command, and control where all can have their decisions informed through a common operational picture. In fact, the Task Force Report promotes this strategy and NIMS. Thus, a police chief or fire chief should be available at the Emergency Operations Center (EOC) or other joint Command Post, and assign an assistant or deputy chief to manage operations at the actual site.

CHAPTER 2: BRIEF HISTORY OF THE NORTHERN ILLINOIS UNIVERSITY SHOOTER AND PARALLELS TO THE VIRGINIA TECH SHOOTER

To some officials and faculty at NIU, the young man responsible for the deadly attack at Cole Hall had been the quintessential student. Described as brilliant, well-liked, respectful, and hard-working by various faculty members and other students, he seemed an unlikely person to commit cold-blooded murder. There were no records of problems with law enforcement or incidents involving threatening behavior on campus during his years as a student at NIU. Only upon deeper investigation did the darker side of Stephen Kazmierczak begin to surface.

As was explained in the Final Report on the Virginia Tech shootings, shooters almost always leave a trail of evidence that they are preparing to commit murder or murder and suicide. Typically, they plan the event in their mind, take action to acquire the necessary weapons or other tools, practice, and then carry out the crime. Seung Hui Cho, the person responsible for the 32 deaths at Virginia Tech in Blacksburg, Virginia, exhibited many "red flags" from the time he was 14 years old. The NIU assailant's background also reflected patterns and a combination of signs that spoke of future trouble. There were clear indications of a deeply disturbed young man whose volatile behavior on occasion caused his family and a few friends to worry about his potential for violence.

The majority of information in this chapter was gathered from an article written by David Vann for Esquire magazine, entitled, "Portrait of the School Shooter as a Young Man," August 2008. It is recognized that this source is not the usual type of source upon which USFA reports rely; however, Vann's article cites acknowledged police sources, and newspaper and magazine stories can often contribute to the body of knowledge about an event.

The man who ended his own life after killing five innocent victims was raised in Elk Grove Village, Illinois, and graduated from the local high school in 1998. He was an above-average student. He had few friends. Some of those he did connect with joined him in destructive activities: lighting chemicals on fire, making Drano bombs, blowing things up, and shooting pellet guns at passing motorists. He was questioned by police when he was in middle school after his mother, having discovered two-liter bottles and the ingredients for Drano bombs in her son's backpack, reported him to authorities. By 11th grade, he was showing friends a business card from the KKK and spray painting swastikas on concrete sewer pipes.

The murderer also suffered from mental illness. Diagnosed as bipolar, he was anxious, depressed, and suffered from insomnia. He would take medication for a while, and then stop taking it because the side effects, including enormous weight gain, were so terrible. Before he graduated from high school, he had been hospitalized six times for suicide attempts or threats. He was teased at school for being "crazy" and suicidal. He told a girlfriend that he wanted to hurt some people.

Many times his parents sought medical help for their son. They also asked the high school to conduct an evaluation of him. Instead, the school gave them a book on how to deal with disabled students. In contrast, the Virginia Tech shooter's high school did conduct an evaluation and developed an individualized evaluation plan (IEP) that accommodated his emotional disability. That assailant's parents also obtained counseling support and medical attention for their son from 8th grade through 11th grade.

In the fall of his senior year of high school, Kazmierczak somehow made it to school after having swallowed an entire bottle of pills. A teacher took him to the nurse's office, where he told the nurse that he wanted to die. Several days later he and a friend were stopped by police for smoking marijuana. His father tried to keep drug dealers away from his son by reporting them to the police. When he graduated, the young man did not go on to college or a job; instead, he moved into a group home for psychiatric patients. Over the next 3 years, he landed in another residential treatment program, obtained and lost several jobs at drug stores and a discount department store, and became inspired to pursue college, enrolling in a few classes at a local community college. Then he stopped taking all medications and enlisted in the Army in September 2001.

For a short time, Army life worked well. The structure helped him function and he was recognized for his ability to shoot without emotional or psychological response. He was proud. Then, in February 2002, he was returned to his hometown, discharged from the Army after they discovered he concealed his mental health history, hallucinations, and suicide attempts. The discharge hit him hard. He enrolled at NIU and by late August 2002, was sharing a dormitory suite with three other students. One of his classes was at Cole Hall, the place where he would murder and wound many students 6 years later.

As was the case with the shooter from Virginia Tech, college roommates of NIU's shooter considered him to be very strange. One of them later told police that their roommate rarely left the dorm room except for classes and to get something to eat. He would not go out, preferring to play video games. A similar profile matched the mass killer at Virginia Tech, whose emotional disability of selective mutism was related to his isolation. The NIU assailant spoke persistently and admiringly of Adolph Hitler, Jeffrey Dahmer, Ted Bundy, and other mass murderers. He examined the methods of the Columbine and Virginia Tech killers. His favorite author was Nietzsche. Even with all of this in the background, the perpetrator did well academically and his social life improved. He earned an A in an extremely tough statistics course, ranking third out of 90 students.

May 2006 arrived and Kazmierczak graduated from NIU, winning a Dean's Award and planned to pursue graduate studies there in the fall of that year. As happened throughout his life, the period of progress was short-lived. His mother died, he began having problems with his girlfriend, and NIU cut back on advanced courses, particularly in the area of criminology, which was to be the focus of his master's degree. He would have to get his advanced degree at the University of Illinois in Champaign, which was a three-hour drive from DeKalb. His favorite professor wrote a letter of recommendation for his application to the University of Illinois. Meanwhile, he stopped going to his graduate classes at NIU, reasoning that the credits would not transfer to the University of Illinois anyway. Instead, he focused on purchasing guns and going to the shooting range. Cho did exactly the same thing during the months immediately before his April 2007 attack, the worst shooting by an individual in the history of the United States. In February 2007, one year before he planned and carried out the multiple fatality shooting and suicide, Kazmierczak bought a Glock .45-caliber handgun. In March, he added a shotgun and another handgun to his collection. Since he had been off his medications for 5 years, the gun purchases were legal.

After the Virginia Tech incident occurred, the NIU shooter became excited and studied everything he could about the Virginia Tech assailant, including where that person bought his guns.[5] He was intrigued with the fact that Cho chained the doors shut at Norris Hall in preparation for the ensuing massacre, and he commented to his new best friend that Cho "obviously planned it out well."

In June 2007, he and his on-again, off-again girlfriend moved to Champaign, Illinois. They got jobs and signed up for graduate classes at the university. With all the major changes, he began to unravel. He was paranoid and anxious, and began having severe mood swings again. At the urging of his girlfriend, he made an appointment with and spoke to a social worker at the University of Illinois health center. A follow up with a psychiatrist at the hospital was made, but before he kept that appointment, and mindful of how psychiatric treatment could impact his ability to purchase guns, he immediately traded in two of his guns and his shotgun for a Sig Sauer .380. The doctor who treated him put him back on medication.

Kazmierczak continued to do well academically, and thus, did not draw any particular attention to his mental and emotional problems. Abruptly, however, he stopped attending classes and began working as a correctional officer in Rockville, Indiana. He liked the training, but the job itself was not what he expected, so he ended that as well. Again he stopped taking his medicine, argued angrily online with friends, and began lining up anonymous, clandestine sexual encounters—a new addiction. (The Virginia Tech murderer also met with a call girl at a hotel a few days before that massacre.)

Right after Christmas 2007, the NIU killer returned to the gun dealer and purchased a new gun and a 12-gauge shotgun. He isolated himself from friends. Early in February 2008, he bought gear from a gun supply store along with a spring-loaded knife—another imitation of the shooter he most admired. Monday morning, February 11, Kazmierczak prepared for the murders he planned. He sawed off the barrel of the shotgun and collected the pistols and ammunition he had hidden. He told his girlfriend he needed to visit his godfather who was ill, and drove three hours to DeKalb, checking into a hotel. Over the next 2 days, he bought books and gifts for his girlfriend. Just like the shooter at the Blacksburg, Virginia campus, NIU's shooter went to the post office and mailed a package, only his was addressed to his girlfriend and not to a major New York City news station. Finally, he took steps to remove electronically-traceable information by removing the SIM card from his phone and the hard drive from his laptop computer. Cho's computer hard drive was never located by authorities either, despite a prolonged search. Figure 1 compares the two shooters and indicates the similarities in their backgrounds and attack.

---

[5] Taken from Report of Proceedings of Debriefing of DeKalb Fire Department Responding to Northern Illinois University on February 14, 2008, taken at DeKalb Fire Station No. 1, on March 7, 2008.

Figure 1: Comparison of the VA Tech and NIU Shooters

Point of Comparison	Virginia Tech	Northern Illinois University
Semester	Spring	Spring
Age	22	27
History of mental illness	Yes	Yes
Felt alienated during adolescence	Yes	Yes
Violent writings/Fascination with violence	Yes	Yes
Talked about or attempted suicide as a teenager	Yes	Yes
Institutionalized/Hospitalized for mental illness/instability	Yes	Yes
Purchased weapons despite mental health history	Yes	Yes
Skipped classes and practiced at shooting range before attack	Yes	Yes
Purchased and used multiple firearms at time of incident	Yes	Yes
Bought a knife and had it with him when he attacked	Yes	Yes
Planned the attack well in advance	Yes	Yes
Parents frustrated with child's emotional problems—sought medical help	Yes	Yes
Passive-aggressive tendencies	Yes	Yes
Denied previous mental health problems when questioned	Yes	Yes
Cruel to animals	No	Yes
Fascinated with weapons as a kid	No	Yes
Had an older, successful sister	Yes	Yes
Was an outsider; considered strange	Yes	Yes
Above average student	Yes	Yes
Admired/Imitated previous mass murderers	Yes	Yes
Had previous incidents involving police	Yes	Yes
Shot victims methodically and unemotionally	Yes	Yes
Committed suicide	Yes	Yes

On Valentine's Day 2008, the perpetrator returned to Cole Hall on the campus of NIU and unemotionally fired a shotgun at students sitting in the front rows of the classroom and at students who tried to flee down the aisles. He knew how to use the shotgun, due to his training from his short stint as a corrections officer. He reloaded and shot three more times with the shotgun (six times altogether), then switched to the handguns to fire 48 more shots. He operated as the Virginia Tech attacker had, calmly, unemotionally, walking up the aisles, aiming methodically, and shooting—again and again. Some victims he shot multiple times. Students begged him to stop, a plea he heeded only when he fired the last bullet into himself. All was silent when police rushed into the auditorium moments after the first 9-1-1 call was received.

CHAPTER 3: THE INCIDENT AND MULTIAGENCY RESPONSE WITH TIMELINE

It was 3:03 p.m. and the class, Geology 104: Introduction to Ocean Sciences, was being held in an auditorium at Cole Hall. One of the students, a 23-year-old geography major sitting near the back of the auditorium was listening to the professor discuss "diatoms and microbiotic animals from the deep sea"[6] when a thin young man carrying a shotgun kicked open a side door at the back of the stage, appeared on the stage to the right of the podium, and walked toward it. The assailant said nothing before firing, first at the professor and then at students sitting in the front row of the large room.

Stunned, a few students at first thought it was a prank. Most, however, immediately realized the deadly assault for what it was and rushed to the aisles to get out of the auditorium. Students were pushing through the rows of desks and seats to get to the aisles while others who had already reached the aisles, raced toward the back of the auditorium. The students were frantic to get out, running around on top of others, screaming. The shooter reloaded the shotgun once (for a total of six shotgun blasts) and then used two separate handguns to continue firing at students as they fled up the aisles to the exit doors. He left the stage and walked up and down both main aisles, firing at students as they fled or remained frozen in their seats. After firing nearly 60 rounds, Kazmierczak then returned to the stage, turned the weapon on himself and died of a gunshot wound to the head.

Most of those who were wounded made it outside and from there dispersed to different locations, primarily the Holmes Student Center and the Neptune residence hall complex. The more seriously wounded were too injured to leave. The horrific attack was over so quickly that when rapidly responding officers arrived at the scene the perpetrator was already dead.

FIRST ACTIONS

Students running out of Cole Hall passed a DeKalb onduty fire shift commander in his vehicle and shouted that there had just been a shooting. The officer immediately radioed this information to the DeKalb 9-1-1 Center. The first recorded call to dispatch logged in at 3:07 p.m. Within a couple minutes, DeKalb Fire and Rescue set up operations at a Staging Area that had been predesignated at the mass casualty incident drill at NIU, and the onduty shift personnel and their vehicles began arriving.

NIU police officers were on campus near the scene and responded in less than 1 minute after the time of the 9-1-1 call. They passed students rushing from the scene, some of whom gave a quick description of the shooter. As police entered Cole Hall, they discovered the dead shooter on the stage, deceased and seriously injured students (including the instructor who was a graduate student), and about a half dozen or more students who were sitting in their seats in shock.

---

[6] "5 Dead in NIU Shooting", Chicago Sun-Times, February 14, 2008

Figure 2: Tactical Police Forces Rush to the Scene

The police officers faced two immediate tasks: rule out the possibility of other shooters so that the fire department's emergency medical responders could be cleared to come on site, and assist in triaging the wounded. The officers approached the seated students and asked them to show their hands and stand up, while simultaneously checking whether any of them had been shot. A couple of the students identified the shooter, indicating toward the stage and saying, "He's the shooter. He did this." Asked if there was anybody else, they responded "No, no, this guy right here." Officers removed students who were not physically wounded from the room and then went to each victim, checking pulses and breathing. Using their emergency medical training, the NIU police officers attended to the wounded, assessing their injuries and providing immediate aid. Other officers were busy establishing first, second, and third perimeters around the scene and responding to the various victim collection points where injured students had dispersed after escaping Cole Hall.

Initial interface between police and fire/EMS senior commanders occurred at Cole Hall. NIU's Department of Public Safety was the lead agency until the shooter was identified and determined to be the only assailant. Then Command shifted to the DeKalb Fire and Rescue Department until all victims had been transported, about 3 hours after the incident began. At that time, the focus shifted to investigation and Incident Command returned to NIU's Department of Public Safety.

At 3:07 p.m., the DeKalb Fire Department requested that two medic units and an engine (a box 10 general alarm) report to the Field House Lot 21—a predesignated Staging Area on campus. The Fire Chief arrived at 3:08 p.m. and established the fire/EMS Command as engines and ambulances began arriving at a Staging Area. Two minutes later, all offduty fire department personnel were called to duty.

Calls to 9-1-1 were coming in from multiple locations, raising concerns that there might be more than one shooter. The DeKalb Assistant Fire Chief conferred immediately with an NIU police lieutenant asking, "Do we know that there's only a single shooter?" The lieutenant radioed others in his department and in about 2 minutes confirmed, "It was just one shooter, and the shooter is dead."[7]

---

[7] Taken from Report of Proceedings of Debriefing of DeKalb Fire Department Responding to Northern Illinois University on February 14, 2008, taken at DeKalb Fire Station No. 1, on March 7, 2008.

Records indicate the time was about 3:15 p.m. when Cole Hall was declared secure. The fire department's Medical Units could then respond to the crime scene from their Staging Area. Fire and EMS Command established a fire officer at each dispersal site. Information on the number of victims and their conditions per triage color was communicated via radio to Command. Command assigned companies from Staging to the dispersal sites. Fire, university police, and convergent responders provided medical assistance to the injured students. There were some medical equipment shortages early in the response. (Since the event, NIU pursued a grant to acquire EMS equipment that would be interoperable with the DeKalb Fire and Rescue Department equipment and which would be prepositioned at sites on campus.)

Between 3:15 and 3:20 p.m., the alarm was upgraded to a second and then a third alarm. At 3:25 p.m., the fire/EMS Command requested two helicopters to respond to the Kishwaukee Community Hospital and at 3:26 p.m.—only 11 minutes after fire/EMS providers were allowed access to the wounded—the first patients from the scene began transport to the hospital. Helicopters were requested, not to transport from the scene to the hospital, but to be available to transport any critically injured victims from the Kishwaukee Community Hospital to a higher level trauma center.

At 3:28 p.m., the alarm again was upgraded, this time to the fifth level, which also included all the resources of the fourth level. By 4:54 p.m., less than 90 minutes after the first victim was transported, the 18th (and last) patient transported arrived at the hospital.

TRIAGE AND TREATMENT

By all measures, the firefighters and police officers who cared for, triaged, and managed the treatment of the wounded did an outstanding job. There were several factors that had to be considered by the responders. They knew that not all the injured were at Cole Hall; those who were ambulatory had fled to other locations on campus. However, the medics did not know how many there were nor the severity of their injuries. What was clear was that the victims who remained in Cole Hall were either already dead or seriously wounded. The medics knew that during the first critical minutes, their resources (personnel, ambulances, etc.) were limited until the additional resources that had been called could get to the Staging Area and receive their assignments. There were three paramedics to deal with nine victims at Cole Hall, all of whom were triaged within 11 minutes. Decisions had to be made about whether to use Herculean efforts and all their resources to try to revive victims who were not breathing and had no pulse or to focus first on several victims who, despite significant injuries, might be saved. Mass casualty protocols establish that the priority should be individuals who have a chance of surviving. When it is clear that a victim will die of their injuries despite massive intervention, and there are other seriously injured victims needing treatment as well, the right priority is to care for those who might live.

The victims were located at seats near the front of the auditorium and in the aisles. Figure 3 shows the locations of the injured and dead victims at Cole Hall. One of the first actions taken was to get the nonwounded students who were in shock to leave Cole Hall so paramedics could access and treat seriously injured victims more easily. Likewise, some of the injured had to be moved so they could be treated and so that other victims could be accessed. The room contained many rows of chairs and desks configured in three sections between which were two aisles. The desks were folded out so it was difficult to move through the rows, especially when victims were on the floor between seats. A decision was made to move a few victims from the seats into the aisles so their wounds could be more effectively treated. After each victim was checked, they were assigned a priority number for transport.

Staff on the first ambulances that arrived thought ahead and brought every backboard they could find and dropped them at Cole Hall while the first patient was being transferred into the ambulance. Medics then could get a headstart on readying the other injured victims for transport so that when subsequent ambulances arrived, patients were very quickly moved from Cole Hall into the vehicles.

Figure 3: Status of Victims at Cole Hall by Order of Transport

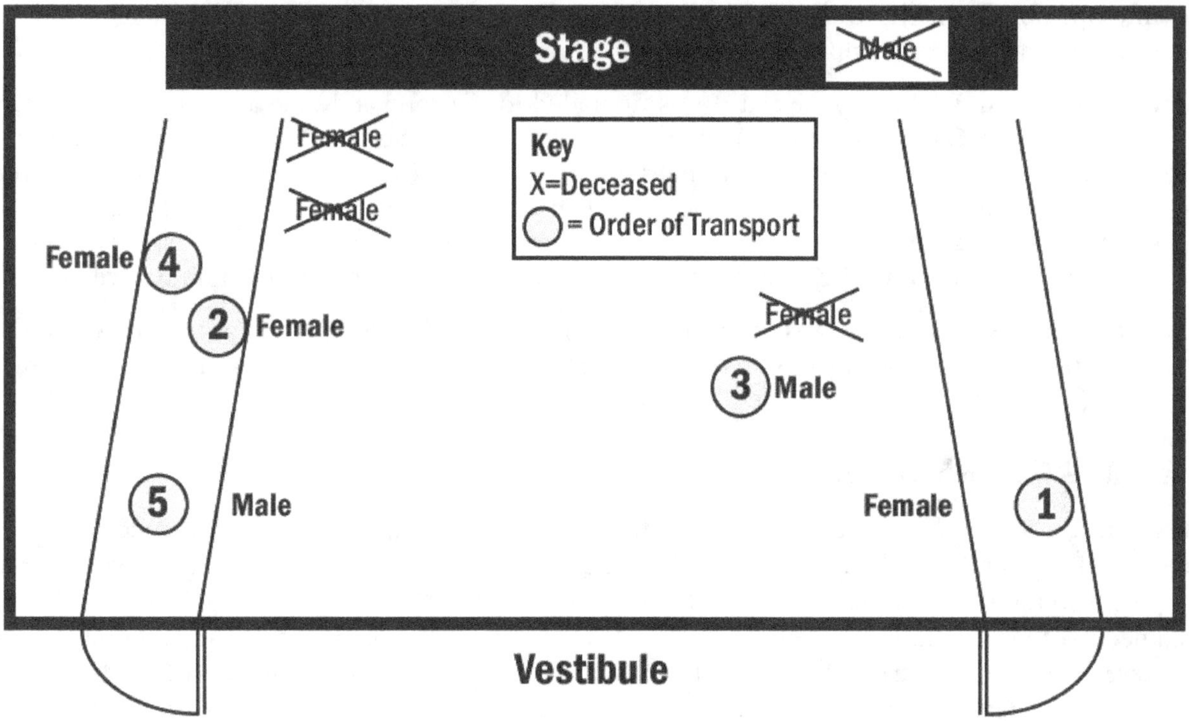

```
┌─────────────────────────────────────────────┐
│                   Stage              X Male │
│                                             │
│              X Female                       │
│                        ┌─────────────────┐  │
│              X Female  │ Key             │  │
│                        │ X=Deceased      │  │
│  Female (4)            │ ○ = Order of    │  │
│                        │     Transport   │  │
│         (2) Female     └─────────────────┘  │
│                              X Female       │
│                                             │
│                           (3) Male          │
│                                             │
│                                             │
│   (5) Male                    Female  (1)   │
│                                             │
└─────────────────────────────────────────────┘
                      Vestibule
```

The medics conducted an effective triage and in so doing, saved lives. The Fire Chief called Kishwaukee Hospital and alerted the Emergency Department that he had a multiple-victim shooting at NIU and that the first six patients the hospital would receive were triaged as red. Updates were provided as the situation developed. The hospital reported later that because they had early notification from the prehospital providers about the number and condition of the victims, they were well prepared to handle the surge in the emergency room.

The first victim was transported at 3:26 p.m. for the short trip to the hospital. That victim had suffered a gunshot wound to the abdomen and chest. Because she was a priority during triage and due to the early alert, Kishwaukee Hospital's surgeons were ready and waiting to take her into surgery the minute the ambulance arrived. She survived. The second patient that was transported had a direct shotgun blast to the chest. She was one of only two conscious patients found in the auditorium. The next patients were taken by ambulance; both suffered direct gunshots to the head, were unconscious, and had labored breathing. The fifth and final patient to depart Cole Hall was the other conscious victim; he had a shoulder injury from the shooting.

The remaining four victims—the shooter and three of his victims—were considered nonviable. Two of the students who were taken to the hospital died later of their lethal injuries.

Figure 4: Paramedics Rush a Victim to a Waiting Ambulance

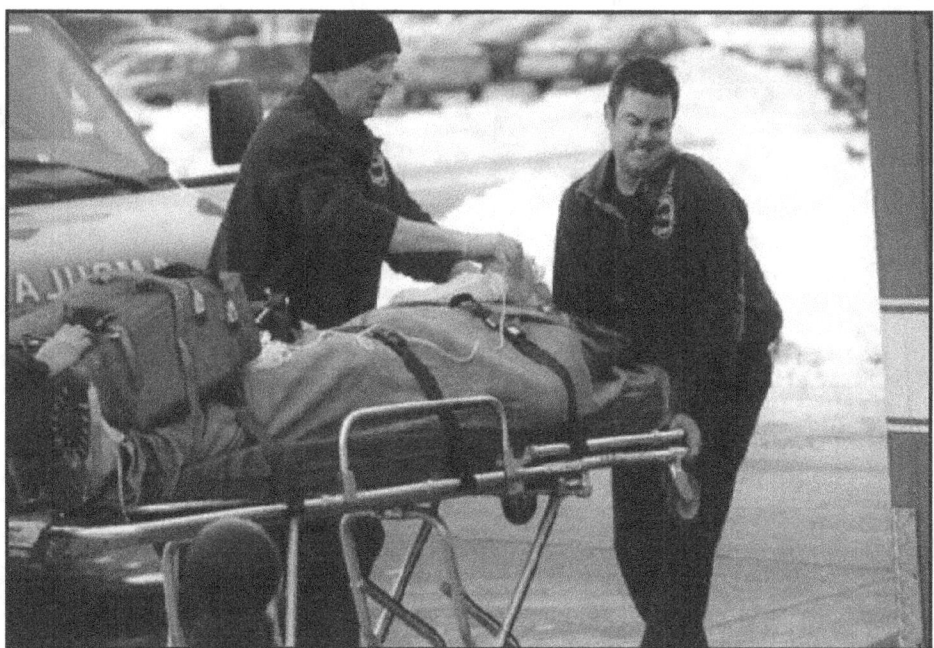

A fire captain/paramedic reported later that they had entered Cole Hall with the intention of using triage tags. They saw the carnage, and knew they did not have a Triage Officer to run the transportation phase and start the triage program. So the three paramedics who were there declared everyone who was still alive as critical, and they proceeded accordingly. Fortunately, all injured victims were going to be taken to the same hospital, making their tracking somewhat easier.

Two ambulances brought equipment to Cole Hall. As noted, Medic 2 and Medic 3 brought extra backboards and supplies, which made a big difference in the ability of the paramedics to manage the casualties. Essentially, they packaged the patients and had them ready so that when the ambulances arrived, the paramedics could quickly slide the backboard onto a stretcher and then out to the waiting ambulance and on to the hospital. The ambulance crews were inside Cole Hall for only a few minutes.

The remainder of the shooting victims dispersed to five locations away from Cole Hall. Three patients went to the Holmes Student Center, six students ran to three different dormitories (Neptune East and Central, and DuSable), and three more students headed directly to the Health Services building. The diagram in Figure 5 shows all six triage locations, including Cole Hall.

Figure 5: Diagram of Triage Locations

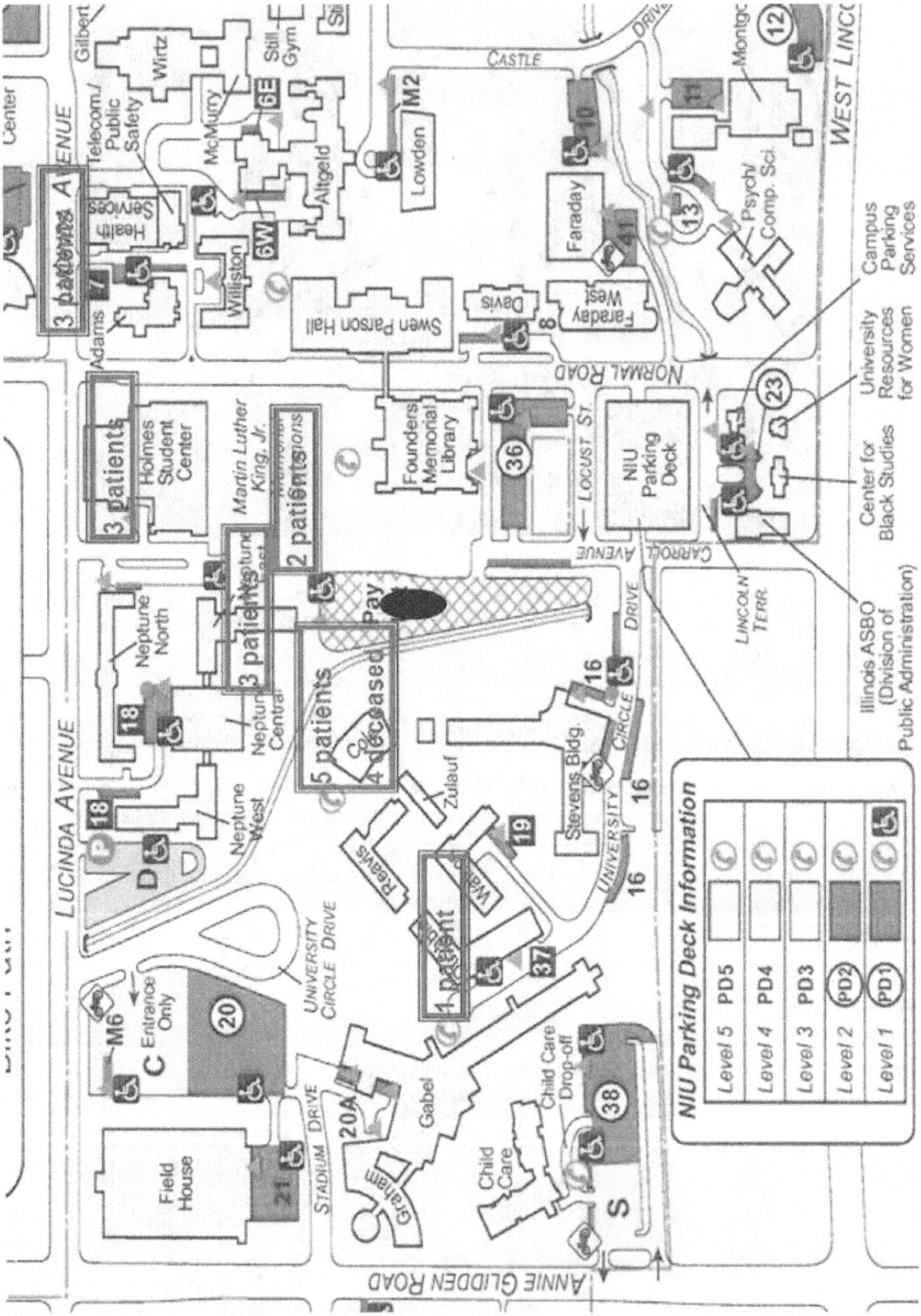

Figure 6: Scene Outside Cole Hall

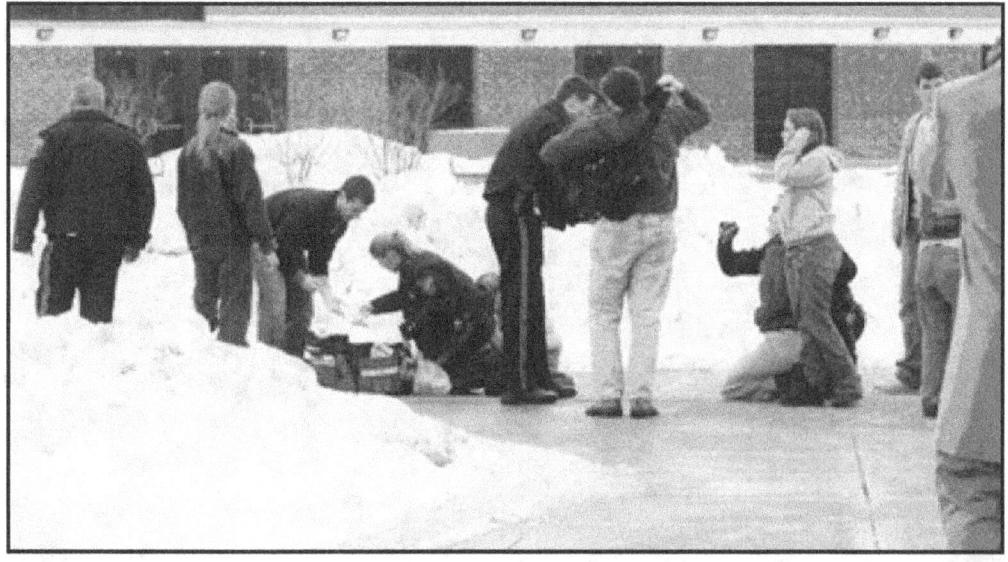

STAGING

DeKalb Fire and Rescue units had trained on mass casualty incident management and had helped direct a drill in October 2007. They carefully reviewed procedures with NIU police and others. After a note was discovered threatening to improve upon the attack strategy from Virginia Tech, first responder officials asked themselves what they would do if the threat were to become reality. The group identified specific areas on campus where responding units would be staged, one for Fire and EMS resources, another for law enforcement, and so forth.

Thus, on February 14, fire commanders knew that rather than respond immediately to the various locations where wounded victims had assembled, they needed to report to Fire and EMS Staging at Lot 21. Part of the rationale for reporting first to a staging site during a mass casualty incident was to ensure that all personnel were safe, that the police knew where they were, and that each unit could be centrally directed to respond once the "all clear" was received from law enforcement, confirming the scene was safe for fire and EMS operations.

After the incident, personnel discussed how Staging had been a secondary concern during the process of developing an MCI plan because Staging had always worked out well in the past, and it did not seem necessary to devote a lot of effort to planning a Staging operation. The NIU incident showed that it is valuable to develop a Staging and Logistics Section of a mass casualty incident plan because logistics proved to be more challenging than was anticipated. Many factors had to be considered in receiving and positioning all the equipment that arrived on the scene.

During the incident there were many pieces of apparatus that responded. Not all of them were used, but at the time of the alarms, the full scope of the incident was not known and they had to be prepared to handle a range of possible contingencies. The area became congested and fire personnel had some difficulties maintaining open lanes for the ambulances. One officer asked for assistance from another firefighter and he took charge of managing the equipment when they arrived at the bottleneck at University Circle.

Figure 7: Fire Department Staging Area

Rather quickly fire officials recognized that Lot 21 was not large enough to handle the number of responding units, and switched the Staging Area from Lot 21 to Lot 20, which was directly across from Lot 21. The shift was made when some pieces of equipment experienced problems turning around and maneuvering in Lot 21. So Staging moved to Lot 20, the closest access point for ambulances to Cole Hall. However, even at that location, due to the layout of the campus, responders had to walk about 150 yards to get to Cole Hall.

Figure 8: Medics Had to Walk a Distance From the Triage Sites to the Ambulances

Once word came that the scene was officially declared secure, medics and ambulances responded from the Staging Area to the six sites where victims were assembled, beginning with Cole Hall where the most seriously wounded were located. The last patient was transported to the hospital at 4:54 p.m. The following chart indicates the victims' locations, transport, and disposition. All victims suffered gunshot wounds.

Figure 9: Status and Disposition of Patients

Patient Location	Treated/Transported by Unit	Patient Disposition at Hospital
1. Cole Hall	Hinckley Ambulance #1851	Transferred to Good Samaritan via helicopter
2. Cole Hall	DeKalb Ambulance Medic #3	Transferred to Good Samaritan via helicopter
3. Holmes Student Center Auditorium	DeKalb Fire Car #4	Treated and discharged
4. Cole Hall	DeKalb Medic #1	Deceased at Kishwaukee Hospital
5. DuSable Hall Snack Area	DeKalb Medic #4	Treated in ER and discharged home
6. Neptune East	Sycamore Ambulance 1-IDA-22	Treated in ER and discharged home
7. South of Neptune East	DeKalb Fire Car #4	Admitted to Kishwaukee Hospital
8. NIU Bookstore–Holmes Student Center	Sycamore Ambulance 1-IDA-22	Transferred to Rockford Memorial via helicopter
9. NIU Health Center	Maple Park Ambulance #1151	Treated in ER and discharged home
10. NIU Bookstore–Holmes Student Center	Genoa-Kingston Rescue Ambulance 1-IDA-11	Transferred to Good Samaritan via ground; Transferred to Univ. of Illinois–Chicago Medical Center
11. Neptune East	Burlington Ambulance #1352	Treated in ER and discharged home
12. Neptune East	DeKalb Ambulance Medic #3	Treated in ER and discharged home
13. Cole Hall	DeKalb Ambulance Medic #2	Surgery–Admitted to Kishwaukee Hospital; Transferred to Northwestern Memorial via helicopter
14. South of Neptune East	DeKalb Fire Car #4	Transferred to St. Anthony's via helicopter
15. NIU Health Center	Rochelle Ambulance #7	Treated in ER and discharged home
16. NIU Health Center	DeKalb Ambulance Medic #2	Treated in ER and discharged home
17. Cole Hall	DeKalb Medic #5	Transferred to St. Anthony's via helicopter, deceased at hospital
18. Cole Hall	(Self transport)	Treated in ER and discharged home
Patient Location	**Treated/Transported by Unit**	**Patient Deceased on Scene**
1. Cole Hall	Not transported	Deceased
2. Cole Hall	Not transported	Deceased
3. Cole Hall	Not transported	Deceased
4. Cole Hall	Not transported	Deceased

As operations began to close down, DeKalb fire officials made a wise decision to hold back several of the approximately 16 ambulances in the event they would be needed to transport some of the patients at Kishwaukee Hospital to other medical facilities, if helicopters could not carry out that mission. The greater alarm ambulances (six) were held to ensure there was reserve capacity. The hospital and the DeKalb Fire Department intend to add this consideration to their plans, ensuring that there are other available transports in case helicopter operations are ruled out for reasons of weather, maintenance, or unavailable due to other prior calls for service.

The difference between the transport situation at NIU and the one at Virginia Tech was that the closest hospital to the campus in Blacksburg, VA, was permitted to divert all other incoming emergency transporters in order to handle the most critically injured victims. Moreover, there were many more victims at Virginia Tech. Due to high winds grounding helicopter operations, all transport from the scene had to be by ground via ambulance or other vehicles to several different hospitals. Kishwaukee Hospital was not allowed to divert because it is the only hospital within 30 miles of another hospital. Kishwaukee is the closest hospital to the NIU campus. All victims were taken first to Kishwaukee, and if necessary, transferred to other hospitals. The other hospitals are at least 30 miles or more from DeKalb and the university.

Law enforcement units were staged at the Wirtz building near the telecommunications and public safety offices of the university. Command was established in the parking lot directly east of Cole Hall. Throughout the late afternoon and evening, law enforcement officers guarded Cole Hall as seen in Figure 10. Figure 11 shows the placement of operations, including the Holmes Student Center where police officials interviewed witnesses, and the boundaries of the hard and soft perimeters around the crime scene.

Figure 10: Law Enforcement Personnel Guard Cole Hall the Evening of the Shooting

USFA-TR-167/August 2009   21

Figure 11: Placement of Operations on Campus

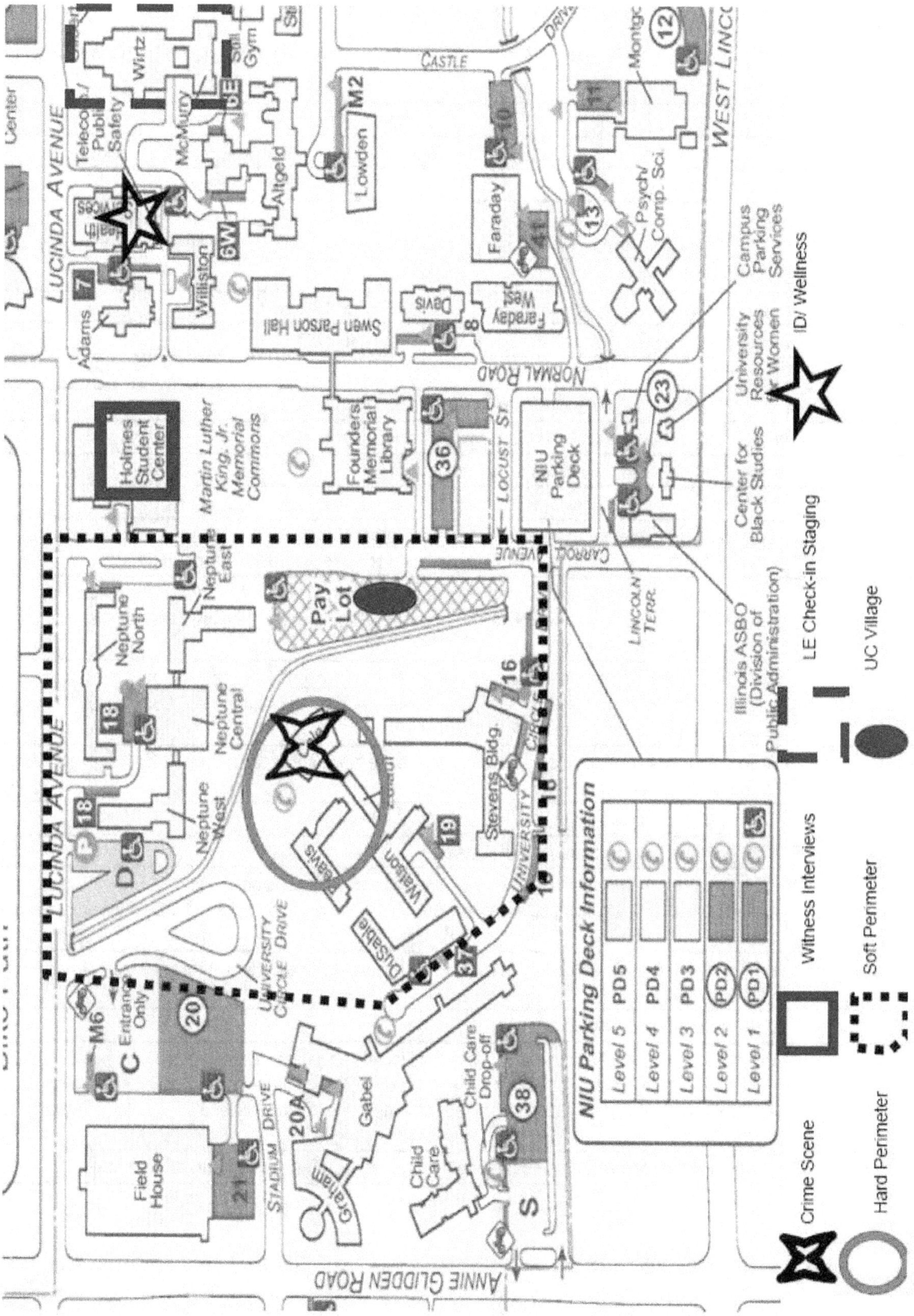

TIMELINE

The following is a combined timeline of events and responses.[8]

Figure 12: Timeline of Events (3 p.m. February 14 – 12 a.m. February 15)

Time	Event
3:03 pm	Shooter appears on stage of Cole Hall classroom and shoots students; then commits suicide.
3:06	**First call to 9-1-1.**
3:07	DeKalb Fire Department Battalion 1 receives report of possible shooting at Cole Hall. Requests Medic 1 and 2 and Engine 1 to respond and stage. Also initiated a Box 10 General Alarm.
3:07	Staging location identified as Field House Lot 21.
3:08	Chief 2 arrives at Cole Hall.
3:08	Fire/EMS Incident Command established at Lot 21 by Battalion 1.
3:09	DeKalb engines and ambulances begin arriving at staging.
3:10	DeKalb recalls all fire department personnel.
3:10	First call to NIU Public Affairs comes from local newspaper reporter.
3:11	Public Affairs calls campus police to confirm.
3:12	Public Affairs calls President for emergency alert authorization.
3:15	**Cole Hall and campus declared secure from further threat. EMS units begin responding, first to Cole Hall from staging. Kishwaukee Hospital receives report of eight victims.**
3:16	Alarm upgraded to Box 10–2nd level. From DeKalb: Engine 3; Rescue 1, Medic 3, Chiefs, and Mass Casualty Trailer. Also four ambulances, a fire chief, one engine and a fire squad, plus another ambulance to fill quarters at Station 1.
3:20	NIU posts an alert on its Web site and sends a blast email telling the campus there was a report of a gunman and giving them directions for safety. The hospital establishes their Incident Command and briefing, and initial assignments are given.
3:21	Alarm upgraded to Box 10–3rd level (one engine, three ambulances, a fire chief, and another ambulance to fill quarters at Station 1).
3:25	Command requests two helicopters to respond to Kishwaukee Hospital.
3:26	**First patients begin transport to hospital.**
3:28	Alarm upgraded to Box 10–5th level (this included all the resources from the 4th level as well)–(two engines, one squad, seven ambulances, two fire chiefs, one mass casualty trailer, and one ambulance to fill quarters at Station 1).
3:30	Hospital decides to use secondary helipad and begins snow removal. NIU crisis team meets in President's suite.
3:38	**First patient is received at hospital.**
3:45	Cell phones jammed at the hospital. Families and friends begin arriving and are taken to the Conference Center at the lower level. Two radiologists report to do wet reads; laboratory assessed O-Negative blood supply and called for more. Phlebotomist reports to help label specimens and send to lab via pneumatic tube system.
4:10	**Campus Police report crime scene has been closed off for the investigation to begin.**

---

[8] Taken from data provided by DeKalb Fire Department; "NIU Shooting Events Timeline," Daily Chronicle, February 15, 2008; the State of Illinois Campus Security Task Force Report to the Governor, April 15, 2008; a presentation given by Kishwaukee Hospital and the DeKalb Fire Department at a Conference of the Illinois College of Emergency Physicians.

4:14	NIU Police publically report that gunman is no longer a threat. NIU announces campus will be closed Friday.
4:15	NIU crisis staff and AVP of NIU Student Affairs at hospital to help with students and families.
4:30	NIU asks all students to contact their parents.
4:39	Public Affairs establishes media staging area in Altgeld Hall.
4:40	Coroner arrives at NIU campus at Cole Hall.
4:45	Victims pronounced dead at Cole Hall by Coroner.
4:53	**Last patient is transported to hospital.**
5:17	Kishwaukee Hospital contacts Coroner of 1 deceased victim from NIU shooting at KCH E.R.
5:30	NIU holds first news conference.
5:45	Hospital contacts NIU Police to confirm names of patients compared to class rosters.
7:00	Emergency Department returns to normal status.
7:30	NIU holds second news conference.
7:55	NIU President confirms that four women and two men are dead and that all casualties were students (the class instructor was a graduate student). Reports basic information about the shooter and indicates that officials are in process of contacting the families.
8:00	Coroner, State Police, and Sheriff's Office arrive at the Emergency Department.
8:00-11:00	Coroner and Sheriff's Detectives meet with families at Kishwaukee Hospital's family center; matching up descriptions of victims with families.
12:00 am February 15	**Kishwaukee Hospital begins receiving fatality victims, which are prepared for viewing by families for positive identification.**

ADDITIONAL RESOURCES

A large number of Federal, State, and local agencies responded to the crisis at NIU. The Communiversity Incident Management Team received support from the Chicago Fire Department Incident Management Team and the State of Illinois Incident Management Team. Other resources that supported the NIU Department of Public Safety and the City of DeKalb Fire Department were:

Local:

- DeKalb City Police—Operations commander and local investigations
- DeKalb County Sheriff's Office
- DeKalb County State's Attorney's Office—Resource support
- Sycamore Police Department—Resource support
- Mutual Aid Fire Departments—Variety of ambulances, special equipment, engines, squads, and chiefs from:

  - Sycamore
  - Malta
  - Maple Park
  - Rochelle
  - Hampshire
  - Burlington
  - Elburn
  - Cortland
  - St. Charles
  - Ogle-Lee
  - Kaneville
  - Sugar Grove
  - North Aurora
  - Somonauk
  - Shabbona
  - Hinckley
  - Genoa-Kinston
  - Waterman

State:

- Illinois State Police–Roving patrol, local investigations, traffic control, intelligence, and security for the governor (during site visit and memorial)
- State of Illinois–Transportable Emergency Communications Systems (ITECS) to support on-scene interoperability and Mobile Command Post

Federal:

- Federal Bureau of Investigation–Local investigations, command post, evidence technician, and crisis management
- Bureau of Alcohol, Tobacco, Firearms, and Explosives–Local investigations

CAMPUS ALERT AND PUBLIC INFORMATION

A member of the local news media was monitoring a police scanner and picked up communications about the shooting at Cole Hall as police were responding. The reporter called the Assistant Vice President for Public Affairs at NIU to confirm there had been a shooting. She tried to reach officials at the NIU Department of Public Safety, but all the officers had left to respond to the scene. She then called the President of the University who had just been notified. The Assistant Vice President for Public Affairs immediately recommended to the President that she activate NIU's Emergency Communications Plan and the President agreed, entrusting her to go forward with the alert to campus and prepare for the media's inquiries. The rapid decision by the President, and the rapid response from Public Affairs meant that the campus received information very quickly on what had happened and what actions they were instructed to take. NIU had a plan for communicating emergency information. They understood the importance of immediate notification through a direct chain of command, rather than waiting until the crisis management team was assembled to discuss the situation in a meeting. When facts are communicated quickly, panic and rumors are controlled.

At 3:20 p.m.—only eight minutes after receiving authorization for an emergency alert—Public Affairs pushed out the first alert on the university's Web site and simultaneously sent broadcast emails and voicemails and recorded a hotline message, all of which stated:

"There has been a report of a possible gunman on campus. Get to a safe area and take precautions until given the all-clear. Avoid the King Commons and all buildings in that vicinity."

Some students could see the drama unfolding from their dormitory windows. A few reported seeing people running and screaming. Looking out some of the windows, resident students saw emergency medical personnel treating wounded victims who had run out of Cole Hall. Intercom systems inside the dormitories alerted students to remain calm and stay in their rooms. A dorm monitor reportedly would not allow students to walk in the hallways and made sure they stayed in their rooms with the doors locked. The dorm monitor demonstrated excellent leadership and responsibility and acted exactly as one would hope during such emergencies.

When Public Affairs posted the first alert they used the "Include" function on the menu to simultaneously post the alert link to the NIU homepage, the Students page, the Faculty/Staff page, and the News & Events section. As the situation evolved and information was updated, they used one page that sequentially documented each update. Every entry was time-stamped, and, based on a tip they had gotten from Virginia Tech, they struck through the text that was updated, but kept it on the page so readers could monitor the progress of updates. A crisis Web site replaced the homepage with special headings: Latest Information, News and Notices, Counseling, Resources and Related Links (e.g., Campus Police, Kishwaukee Community Hospital, Psychological Services Center, etc.) and the hotline numbers. A video of the first press conference was later included. Links to condolences, vigils, and community response were added on the second day.

Throughout that day and into the next, NIU held multiple press conferences. The first was held at 5:30 p.m. on February 14. It covered the basic facts which were confirmed at that time and announced which law enforcement agencies were involved: University Police, DeKalb, Sycamore, FBI, ATF, and State Police. NIU told the media to watch the NIU Web site for updates and future news conference scheduling. That press conference and all subsequent press conferences were streamed live on the Web site, which was heavily accessed by parents, alumni, and others.

The university's Public Affairs Office was prepared in many ways. Providentially, the office had reached out to the Public Affairs Office at Virginia Tech several months after that tragedy to ask what advice they might have on preparing for and dealing with a mass casualty incident. The Virginia Tech Public Information Officer was generous with his time and shared information about the lessons they had learned. The Assistant Vice President for Public Affairs was included on a committee charged by the NIU President with a line-by-line review of the final report produced by the staff and members of the Virginia Tech Review Panel. NIU's Virginia Tech Review Committee used that report as a template for considering additions and changes to their own plans.

In another proactive step toward preparedness, the Public Affairs Office at NIU monitored Virginia Tech's website for a long time after the slayings at that campus, and they remained current with the memorials, events, and details about the incident so they could be informed of the issues that Virginia Tech faced which NIU wanted to capture for their own planning.

One of Virginia Tech's recommendations to NIU during their conversations about lessons learned was that NIU should secure additional servers to handle the communications surge that arises when any major incident occurs. NIU fortunately did buy and install six additional servers, otherwise, their site almost certainly would have crashed (at one point they had 14 million "hits"). The Public Affairs Office was deluged with calls from parents seeking information about their children, foreign consulates wanting to know if any of their students were affected, and the media. All the local media indicated they were on their way to the campus and the sound of media helicopters overhead followed shortly.

At 3:30 p.m., only 20 minutes after the NIU President learned of the shooting, the crisis team assembled in the President's conference room which became a de facto emergency operations center (EOC). The NIU crisis team consists of:

• President	• Assistant Vice President for Public Affairs
• Provost	• Vice President for Government Relations
• Chief Operating Officer	• Chief Legal Counsel
• Vice President for Student Affairs	• Executive Director, Community Relations
• Assistant Vice President for Information Technology	• Police Chief

The police chief remained on the scene at Cole Hall, rather than reporting to the university's EOC at Altgeld Hall, so he could communicate basic facts via radio. The crisis team decided to cancel classes until further notice, acted to make counseling available, scheduled a news conference for 5:30 p.m. and established student/parent hotlines. University officials had experience setting up and using a hotline because the previous semester they had needed to activate a hotline operation twice: during a flood in late August on the day before fall classes were to have begun, and in December, when a graffiti threat was discovered in a residence hall bathroom. Also, hotlines were part of the mass casualty drill they had practiced during homecoming. The Division of Student Affairs staffed the hotlines and had at their disposal the most up-to-date information about the incident and what was being released to the public.

Altgeld Hall, the main administration center of NIU, was established as the media center. Public Affairs directed operations of the media center, including opening two conference rooms for the media, provided wireless Internet access, and began collecting media names, e-mail addresses, and other contact information, and generally seeing to the reporters' needs. It was recognized the President needed to be highly visible, and as much information as possible needed to be released as quickly as possible. The media had inundated the campus with reporters, vehicles, and equipment. They had an edge on the university in reporting some details because the media had dozens of reporters near the scene and all over campus. One of the lessons NIU learned was that all public emergency communications should direct people to the website for details and updates. Linking the NIU Web site to that of Kishwaukee Hospital also proved helpful to families and friends seeking information about loved ones.

A series of news conferences was carried out. Updates on the status of the victims and the university's activities were explained to the media. From their counterparts at Virginia Tech, NIU Public Affairs officials also picked up helpful tips for the Web site, such as how to "go dark" with the banners and background so that what one sees immediately is the emergency information—one does not have to search through extraneous text on the homepage to locate that information. Key timeline entries for the public information that was communicated are included in the earlier timeline table.

Figure 13: NIU President, Dr. Peters, at the Second News Conference

Throughout the course of the afternoon and evening, NIU Public Affairs staff continued to feed new information to key stakeholders, wrote and disseminated public updates, handled media calls, and gathered, created, and distributed photos, maps, diagrams and so forth. Staff monitored news coverage, kept track of social media and blogs, and constantly updated the NIU Web site.

The NIU Public Affairs Office drew several conclusions from their experience. They are:

- Have a plan. The value is in the process more so than the product.
- Practice the plan and continually revise and update it.
- Have backups for all key positions.
- Involve colleagues from information technology and Student Affairs—the new electronic media reigns.
- Identify and train spokespersons; be selective in their deployment.
- Nurture good media relations before crisis; treat media equitably during the crisis.
- Keep key stakeholders informed.
- Reach out and accept help.
- Treat media as partners, not enemies.
- Anticipate and prepare for "news triggers" (first day of classes, anniversary of the shooting, etc.).
- Keep victims and families in mind.
- Speak from the heart.

CHAPTER 4: HOSPITAL AND CORONER'S OFFICE RESPONSE

Kishwaukee Community Hospital opened a new, state-of-the-art facility in October 2007. It is an affiliate hospital in the Illinois Trauma System—not a Level 1 Trauma Center, but nevertheless can handle, and does handle trauma patients within the Emergency Department. The hospital has 15 beds in the ED, 13 private rooms, and two trauma bays, though on the day of the incident, the other rooms were used to handle traumas as well. The hospital is only 5 minutes away from the Northern Illinois University campus. One of the doctors, who was managing the Emergency Department at Kishwaukee Hospital the afternoon of February 14, participated in a comprehensive debriefing on February 28, 2008. The debriefing was held at the hospital and attended by the fire department EMS staff who responded the day of the shooting. The details contained in this section are drawn heavily from the transcript of that debriefing and from the after action presentation the hospital representatives gave at a conference later that year.

At the time of the shooting there were seven nurses, two EMTs, a unit clerk, and two emergency physicians on duty. An additional nurse and physician were en route to the hospital to open their Fast Track area at 4 p.m. Notification to the hospital staff about the shooting came from multiple sources: the DeKalb Fire Department, family members who worked at NIU or who were police or fire personnel.

Once notified of the shooting, the hospital quickly formed a team of doctors, surgeons, nurses, and other technical staff to handle the incoming patients. The doctor in charge assembled five doctors, four general surgeons, and six orthopedic surgeons. Other medical staff, nurses, radiologists, laboratory personnel, and so forth supported the surge requirements to treat the wounded.

Kishwaukee Hospital activated the Illinois Regional Hospital Coordinating Center system through the region's designated center at Rockford Memorial Hospital located about 40 miles northeast of Kishwaukee. The Illinois Department of Public Health has established 11 regional EMS and trauma areas and designated a lead hospital in each region that is responsible for coordinating hospital support during major emergencies and for managing regional preparedness planning. That hospital serves the role of disaster assistance coordinator for other hospitals in the region that are confronted with a medical emergency situation which presents a challenge to the hospital's normal resources and capabilities.

Rockford Memorial Hospital, as the lead hospital for the region in which Kishwaukee is located, helped to manage additional medical-related resources. The lead hospitals in the system are responsible for finding extra ambulances at the required levels (basic, advanced life support, etc.), for obtaining helicopters for air medical transfers, and for other medically-related assistance. They draw resources from other hospitals and transport services in the region and coordinate with the Illinois Emergency Management Agency. The designated hospital establishes a command site and remotely manages support from that location. The afternoon and evening of the shooting, Rockford worked to obtain helicopters from the east and the south because one of the helicopters normally used was down for maintenance, and weather problems to the north affected the availability of helicopters in that area.

The Illinois Department of Public Health provided immediate support. Law enforcement officers arrived to secure the emergency room. This was a wise move to protect both hospital emergency personnel and the patients. In these types of situations—mass casualties as a result of criminal action—it is important to protect against the possibility that whatever happened at the scene could transfer into the hospital environment. Medical staff needs to focus on treating the wounded without having to worry about intruders or people bent on retaliation finding their way into treatment areas.

There were many preparations that had to be made. Hospital personnel had to consider relocating some of the nine patients who were already in the emergency department. Rooms had to be made available and prepared with a trauma flow sheet and other forms. They needed extra crash carts, chest tube set-ups, and blood. They had only one available helicopter and called Rockford Memorial for more.

Once patients began arriving, the hospital had difficulty with patient identification. Because the scene had been secured so quickly and the victims moved out so fast, no triage tags were used. The students' belongings were scattered, and information that would typically identify them was in their backpacks back at Cole Hall.

A mass casualty incident with gunshot wound victims is a complicated and problematic scenario. As noted by the doctor during the debriefing, when medical staff begins treating gunshot-wound victims, the ballistics of the guns that were used is not usually known, so the extent of internal injury is difficult to assess at the outset. The location of the entry wound is only the starting point. It may appear as though the injury is minor, but in reality the bullet may have taken a direction that affected major organs, veins, or arteries. Bullets can move anywhere. In fact there were several victims who at first appeared to have sustained only minor injuries, but who in fact, were critical. Hospital staff knew that the vital signs with young victims can change quite rapidly; they can be stable and maintain that condition for a long time and then suddenly drop to critical levels.

Kishwaukee had two radiologists who came down to the emergency room immediately and conducted real time positron emission tomography (PET) scans. As scans were taken, the radiologists read the results which aided in determining who needed surgery and in what priority order. The information also helped with decisions concerning which victims needed to be transferred to other hospitals (Kishwaukee Community Hospital is not a trauma center).

Later, the hospital said they were pleased with how well their system and response plan for a mass casualty incident worked. They noted that the disaster exercise from the previous year, when the medical response system was tested for a mass casualty event, had contributed to the plan's success when made operational and it saved lives. That plan was the basis for their response and they flexed it to fit the situation where needed.

Three of the injured victims were critical and had to be resuscitated. One had a tension pneumothorax, another tension hemothorax, and a third who presented with a blood pressure of 60 who had hemoperitoneum and cardiac involvement. All were saved.

Seventeen individuals were transported to Kishwaukee Community Hospital. From there, eight victims were taken to other hospitals in the region. Two of the 17 who were transported subsequently died—one at Kishwaukee Hospital and another at St. Anthony's Hospital. Figure 15 shows the location of the transfer hospitals.

On Friday morning, February 15, the emergency room director addressed the media at a press conference.

Figure 14: Dr. Michael Kulisz Speaks to the Media at Kishwaukee Community Hospital

Figure 15: Hospitals Involved in Treating the Victims

*One patient underwent surgery at both hospitals and is counted at both locations.

ARRIVING FAMILIES AND INFORMATION

Most of the NIU students contacted their families after the slayings to let them know they were safe. In fact, the 4:30 p.m. public communication from NIU urged students to call their parents if they had not already done so. Since the vast majority of NIU students are from communities within a 100-mile radius of DeKalb, family members of the students who were killed and wounded began arriving shortly after the incident and went directly to the hospital. The hospital's communications system was inundated with calls, and the attending physicians in the Emergency Room were contending with multiple communication issues. They were dealing with EMS, the police, the State, families, in-house physicians, and trauma surgeons. Some of the wounded students wanted to hand their phones over to the doctors so the doctors could brief a parent or a sibling about the student's condition, all while orders were flying back and forth and treatment was being administered.

Emergency room doctors at Kishwaukee hospital allowed those among the most critically wounded who could call their families from their cell phones to do so before they were transported to other hospitals via helicopter. Otherwise, hospital personnel were constrained by the HIPAA law which prevents health practitioners from divulging names. Thus, when calls came in asking for information on the whereabouts and condition of students who were being treated, nurses and administrators could not provide that information, mainly because over the phone, they could not be certain with whom they were dealing. In some cases. family members who could be there in person were able to get in to see their children.

The hospital set aside their Conference Room for arriving family members and also separate rooms away from media. Snacks and beverages were provided while family members and friends waited for news. NIU Crisis Staff arrived to assist the social workers at the hospital. Members of the clergy reported to the area to be available to the families. State Police, Sheriff's Office, and the Coroner arrived between 8:00 p.m. to 11:00 p.m. to work with the families on matters related to identification of the wounded and the deceased.

Three of the victims at Cole Hall died at the scene, as did the shooter when he committed suicide. Two other students died of lethal wounds at hospitals: one at Kishwaukee and one who was transported to St. Anthony's Hospital. Cole Hall was secured by police. The DeKalb County Coroner's Office, State Police, and the FBI began moving the human remains to the hospital at midnight. The last fatality was transported out of Cole Hall at 12:20 a.m., February 15.

At the hospital, personnel from the Coroner's Office and State Police met with families to collect identifying information, e.g., height, weight, hair color, and markings (tattoos, moles, etc). Complete body x-rays were done for all of the fatalities, the remains cleaned and draped, and positive identification made by families. The Coroner then cleaned and draped the remains, the families were then permitted to make a positive identification. At St. Anthony's Medical Center, the Coroner in that county identified that victim, working with the victim's family.

The hospital administrators and the Coroner worked together to make sure families who wanted rooms at local hotels were able to get them and transportation was provided for them. The Coroner exchanged information with the families so he could contact them the following day. If the families needed information about a funeral home, the agencies that were staffing the hospital's family area made the connections and arrangements.

The next day, maintaining a chain of custody, the remains of the deceased were transferred to the Coroner's Office in Sycamore, Illinois, for autopsies. The Coroner had checked with the families as a standard protocol regarding whether they wanted to donate tissue. As each autopsy was completed, the Coroner called the family and personally talked to them about his findings, providing answers to their questions and facilitating a level of closure for them.

SECOND WAVE OF PATIENTS

After any major incident, typically there is a second wave of patients who present themselves to clinics or the hospital—individuals who usually have minor injuries, or who are feeling stressed and ill after the initial shock of the event passes. Such was the case after the NIU shooting. Kishwaukee Community Hospital's social workers attended to several individuals who sought treatment for minor injuries, the effects of anxiety and stress, and so forth. However, the number was lower than expected. This was attributed to many students leaving the college and going home until classes resumed on February 25.

CHAPTER 5: AFTERMATH AND HEALING THE WOUNDS

Makeshift memorials to the slain and injured students began appearing on NIU's campus within hours of the slayings. Flowers, cards, teddy bears, and messages of remembrance were some of the ways that mourners used to share their feelings about the tragedy and to join others in tribute to the victims. People lit candles at a midnight vigil and huddled together for support. On February 19, the Tuesday after the shooting, parents of the slain students were escorted by the NIU Police Chief to view the memorials on campus and to enter Cole Hall if they wished.

Figure 16: Mourners Participate in Candlelight Vigil

The Director of NIU's counseling and student development center oversaw counseling support services. At the hospital, NIU crisis staff went directly to the Conference Center at the lower level of the hospital where a family assistance center had been established. The university began planning a memorial service to honor the victims of the shooting. In a statement announcing the service, the NIU President set the tone for the memorial as the beginning of a journey toward healing that would take time. He stated, "This memorial will initiate a set of activities and services aimed at community recovery that will continue throughout the semester."

Figure 17: Crosses Erected Near the Site of the Shooting

One week after the crisis, NIU and the surrounding community observed 5 minutes of silence at 3:06 p.m. while bells at the Holmes Student Center and various churches in the area chimed for 5 minutes—1 minute for each of the students who died.

On February 24, 10 days after the incident, 12,000 people assembled in the NIU Convocation Center to honor the slain students, to hear words of encouragement, and to find emotional support. The NIU President spoke of hope and the bridges that existed among the attendees and the world. State and Federal officials also spoke about the university community's strength and ability to move beyond the sadness of that day. The memorial testified to the enormous support that all who were affected could draw upon while dealing with the emotional debris that such senseless killings leave behind. NIU's combined chamber and concert choirs sang at the memorial, helping all assembled to say goodbye and move on to the future.

Figure 18: A View of the Packed Convocation Center During Memorial Service

NIU also coordinated nearly 300 counselors who volunteered to assist at NIU. They attended the memorial service at the Convocation Center and then were positioned in classrooms, residence halls, and department offices to help returning students and faculty cope with the transition and sense of loss once classes resumed. The university hoped that their presence would contribute to healing and a sense of unity within the campus community.

CRITICAL INCIDENT STRESS

Few incidents are as traumatic to responders and witnesses as a mass shooting that involves innocent young people. And in small communities, there is a greater likelihood that fire and law enforcement personnel may be acquainted with one or more of the victims. Certainly for the campus community as a whole, the victims were part of the NIU family which experienced great shock over losing fellow students and friends.

The City of DeKalb Fire Department took several steps to mitigate the stress among those who responded to the shooting. Early in the event, the fire department issued a department recall which brought a substantial response from off-duty crews and mutual aid companies. Later during the evening of February 14, the Fire Chief brought the shift that was on duty (and that responded to the incident) to Station 1, took them off the apparatus and relieved them of their responsibilities. The department brought in a critical incident stress management team which conducted a diffusing session with the firefighters, after which they were permitted to go home. One of the concerns the Fire Chief had was that someone from the college might call in a false alarm about another shooter, and that such a call could magnify the stress the responders already were feeling from having been part of the horrific scene at NIU that afternoon. The shift was a tight-knit group and that helped them work through the issues they faced. The day after the shooting, members of the shift went to lunch together—an event that was supported by the DeKalb Firefighters Local #1236. Other forms of support were provided throughout the fire department.

On the following Tuesday night, February 19, a formal debriefing was held for all responders: firefighters/medics, police officers, dispatchers, mutual aid companies and any others who were part of the response at NIU. The debriefing was conducted at the Convocation Center on campus. It is important to note that dispatchers were included in the debriefing. Sometimes these personnel are overlooked, but, they, as much as any responder, are under great stress as they process and prioritize all the calls for assistance and the multiple alarms. Often, they do not receive much feedback on outcomes, but they should, so that they can feel a sense of completion about their role and contribution to the overall management of resources and response.

Over the next days and weeks, there were informal conversations as personnel processed the experience with their peers. The fire department also reminded all personnel about other options and resources that were available if they needed to talk one-on-one with a professional counselor or doctor. The fire department organized and facilitated two specific technical debriefings as well. These were set up to clarify details about the response and to discuss what was successful and why, as well as what changes to procedures and strategies they might want to make.

On February 28, there was a comprehensive medical debriefing session held at the Kishwaukee Community Hospital Training Center with the emergency medical responders, the doctor who directs the Kishwaukee Emergency Department, the DeKalb City Manager, and an NIU Police Lieutenant who serves as the emergency management coordinator for the campus. The President of the Illinois Fire Chiefs, and the Chairman of the EMS Committee of the Illinois Fire Chiefs also sat in on the debriefing to learn more about how the response worked and why it was so successful, and what ideas should be carried forward to improve future operations.

About one week following the medical debriefing, the fire department directed a review of the fire response and management of the incident, including coordination with NIU police. Fire department mutual aid companies were represented along with an NIU Police lieutenant. The goal was to confirm the timeline and document key aspects of communications, staging, strategies, and command. They also discussed some problems that developed and how they were solved. During this meeting, the fire department acknowledged two letters of support that had been sent: one from the Virginia Tech Rescue Squad which expressed empathy for what the responders and the university were going through and another from a former officer with the Champaign, Illinois, Fire Department who had coordinated the unified command classes that many of the responders had attended. He also had promoted the concept of unified command to the State of Illinois. That training led to changes in emergency management which responders proved they had mastered. They were prepared.

A group of DeKalb county volunteers specially trained in critical incident stress services provided post-incident debriefings for first responders. Local churches and religious organizations provided a number of "Care for the Caregiver" workshops for all who were affected, bringing in a nationally known speaker. The NIU Employee Assistance Program provided debriefings for all police officers who were first responders; this group received two additional post-incident debriefing sessions. Some of the students, who were among first responders, were also offered services.

All faculty and staff were offered group and individual sessions, and two-thirds of the university faculty/staff population attended one or more group sessions within a week after the event. Hundreds more took advantage of individual sessions to help in the response. In addition, further group and individual sessions were provided for faculty and staff who were involved in or witnessed the events at close range. During the semester, three recovery workshops were offered for faculty and staff, and the university invited the faculty and staff population to special workshops held at the return to classes in the fall. Additional workshops with recovery themes were made available for both semesters of 2009. NIU EAP identified faculty and staff who had been closely affected by the events, and provided personal outreach and follow-up discussions with faculty ranging from those who had known the shooter, those who had provided immediate first aid and support (e.g., Founders Library, English Department, Women's Studies, Neptune Hall, Holmes Student Center, Communication, Cole Hall basement, carpenters, Sociology and Geology). EAP worked with deans, chairs and administrators on what their faculty and staff needed in the aftermath.

NIU's Counseling and Student Development Center offered workshops and groups for students in the months after February 14, and further workshops on the return to classes in the fall. All of the campus counseling centers and training sites report that counseling centers have been functioning at or near capacity since February 14, and this trend continues.

CHAPTER 6: FINDINGS AND LESSONS

No two disasters are exactly alike. Even though the type of disaster may be familiar territory—for example, hurricane, flood, or shooting, each one reveals something new about preparedness and response which adds to the collective knowledge about what needs to be considered, what works, and what may not work. A mass murder at a university is still a rare event; however, when it occurs it draws significant national and international interest in the lessons that can be drawn from the tragedy. In the case of the February 14, 2008, slayings at NIU, there are issues that the DeKalb Fire Department, the Coroner's Office, the NIU Public Affairs Office, and Kishwaukee Hospital have identified and shared. In addition, USFA has drawn several other conclusions upon an objective review of the incident.

To the community's great credit, they had been proactive in preparing for a possible incident on campus, including a mass casualty incident, for many months before the tragedy. And they give significant credit to Virginia Tech for unselfishly sharing information on a wide variety of topics, including how to handle the public information part of critical incident planning several months after that university suffered their terrible loss. The detailed after action analysis and report on the Virginia Tech incident, which was produced by a special panel and staff appointed by the Governor of Virginia, was mentioned by nearly every office or department in their debriefings and conversations as having been a critical source of information for all involved in public safety, emergency medicine, emergency management, and recovery at the university and the related first responder organizations. What this proves is that incident critiquing and sharing the results with others succeeds in accomplishing two things: they produce something good out of a wrenching experience, and they make a difference in helping other jurisdictions enhance their strategies and tactics for improved outcomes based on experience driven by actual events.

The State of Illinois's campus security initiatives also made a difference. Grant dollars provided radios and special training and benefited many of those who responded to the crisis. Police, fire, and emergency medical responders had developed a strong team. That team practiced together, planned together, and responded to incidents together in the months and years which preceded the Cole Hall slaying. "Relationships, it's all about relationships" was a commonly voiced conclusion. That same reaction has been documented in USFA's report on the response to the Minneapolis I-35W Bridge Collapse[9]—another multi-agency, intergovernmental response situation.

There are many specific lessons that can be captured from reviewing the NIU incident. These lessons are detailed in the following paragraphs, and include confirmations of known good practices and discussions of actions or gaps that were problematic. It should be noted that the lessons cover many aspects of the response, except that USFA did not receive any direct information from the NIU Department of Public Safety, despite attempts to obtain that information. What is mentioned concerning police operations was derived from hospital, fire and EMS, and other after action reports that included some facts about police response. USFA also collected police department data from the NIU Web site, the annual Clery report, and newspaper articles.

---

[9] I-35 Bridge Collapse and Response USFA-TR-166, March 2008.

1. **Planning Together**—As noted throughout this report, the preparedness planning and training that was shared among responder organizations contributed to good working relationships, trust, and the rapid triage, treatment, and transport of the victims.

2. **Police EMT's**—The NIU police officers first on the scene at Cole Hall used their emergency medical training to great advantage. This training is valuable and noteworthy.

3. **Notification of Coroner**—Too much time elapsed between the murders and the formal notification of the Coroner. Generally, such notification is made by law enforcement; however in this case, the hospital contacted the Coroner.

4. **Public Warning and Information**—NIU's President immediately authorized activation of the Emergency Communications Plan. The series of messages that were broadcast were clear, described the situation, and told people what to do. The Public Affairs Office used the NIU website to great advantage. A hotline staffed by Student Affairs handled more than 10,000 parents and family calls in the first 48 hours. News conferences were well managed and had clear missions. Care was taken to brief all media at the same time.

5. **Fire Staging**—Fire apparatus and ambulances did exactly what they had established and practiced—that is—they reported to a pre-designated location close to classrooms rather than arrive at the various sites where victims had dispersed. The trust that existed between police and fire added credibility to early assurance that there was only one shooter. Fire personnel knew that police would not put unarmed firefighters in harm's way.

   The fire department showed good flexibility when they changed to the staging location to a site across from the original site once it became clear that they needed more room and a wider turning radius.

6. **Fire/EMS Operations**—Fire department paramedics and EMTs, along with the police EMTs saved lives. Victims were assessed and handled per proper procedures for mass casualties. An ambulance crew was prudent in collecting and delivering extra backboards and supplies so that injured victims could be moved quickly into awaiting ambulances. Fire ICS was followed.

7. **Hospital Response**—Kishwaukee Hospital had an emergency plan and set it in motion when they were alerted to the shooting. The Emergency Department acted quickly to clear and prepare space and secure additional equipment and personnel. The hospital was secured by law enforcement quickly, a good step to take until they could be certain the shooting was not gang-related.

8. **Coroner's Response**—The Coroner came to the hospital once notified and worked with the families to collect information that would help confirm the identity of the deceased. The families were dealt with sympathetically. They were permitted to view their children as soon as procedures and arrangements could make that possible, The Coroner followed up with each family personally to discuss the procedures that would be taking place by their office. With the other care providers at the hospital and public safety personnel, the Corner ensured that families had provisions for the night and transportation. The following day the Coroner contacted each family to discuss the findings of the preliminary autopsies.

9. **Assistance to Victim's Families**—NIU's Emergency Operations Plan (EOP) includes provisions for assisting victims' families in the immediate aftermath of a tragedy. Immediately following the shooting, NIU's Division of Student Affairs appointed staff liaisons to each of the families of the deceased and to each injured student and the student's family. In the earliest hours following

the shooting, liaisons were dispatched to Kishwaukee Community Hospital where family members had congregated and awaited word on the status of their loved ones. The hospital arranged for a family assistance area, and that area was suitable for the purpose insofar as it was away from media in the hospital and was located where access could be controlled. Social services, faith leaders, counselors from the university and hospital, the Coroner, and State Police all worked on behalf of the families at the family assistance area the afternoon and evening of the shooting.

Several days after the shooting, NIU established the Office of Support and Advocacy (OSA) to handle the longer-term needs of victims and families. OSA provided central resources and one referral point for those students, families, faculty, and staff most directly impacted by the February 14, 2008, tragedy. OSA services include:

- Support and referrals to enhance academic and life success;
- Academic counseling and support services;
- Individual, group, and family counseling services;
- Appropriate accommodations for academic success;
- Victim assistance and advocacy;
- Work with on- and off-campus departments and agencies to identify and provide specific services;
- Initiate structured activities and communication processes to support networking and mutual support efforts; and
- Provide leadership to the campus community as a strategy for developing and maintaining a campus climate sensitive to the experience of those most impacted by the tragedy.

In addition to these services, OSA worked with NIU's Legal Services department to facilitate the many services available to victims of crime. For example, there is a Federal[10] fund that is implemented at the local level through either the prosecutor's office or the police department which compensates victims and their families for a variety of expenses related to crime. Travel and funeral expenses, for example, are sometimes included. There is a set death benefit to which families are entitled. Victim assistance programs can help families cope with the crime in various ways, and those programs were and continue to be used by the victims and families.

10. **Critical Incident Stress**–The DeKalb Fire Department took great care to ensure that responders to the shooting were relieved of their duties as soon as was practical the evening of the shooting. They were given a chance to talk and defuse in an informal session led by a professional in this field. The department organized a full debriefing shortly thereafter where all personnel who were involved could process the event. Personnel also were informed of additional services that were available, and to be aware of the danger signs of stress.

NIU acted quickly to establish a variety of support services and counseling venues for all who sought help. Families, students, faculty, and staff were provided many opportunities for counseling and support in different settings over the weeks and months following the shooting, and into the following academic year.

---

[10] http://www.ncvc.org/ncvc/Main.aspx; http://www.ojp.usdoj.gov/ovc/

11. **Debriefings and Post Incident Critiques**–During the research for this report, several transcripts of debriefings and various after action presentations were made available. The hospital, the fire department, the Public Affairs Office, and the Coroner all are to be congratulated for their excellence in documenting their actions and decisions throughout the emergency and for making this information available. They performed a great service to their employees and to others by doing so—their information was vital to this report.

12. **Recovery**–The scheduled chiming of the bells, the memorial service, and the counseling support on campus all contributed significantly to the sharing of grief, the honoring of the deceased and injured victims, and the initiation of healing.

Figure 19: A Demonstration of Hope for the Future

www.ingramcontent.com/pod-product-compliance
Lightning Source LLC
Chambersburg PA
CBHW081625170526
45166CB00009B/3102